Standard Grade | Credit

Chemistry

Credit Level 2001

Credit Level 2002

Credit Level 2003

Credit Level 2004

Credit Level 2005

Leckie × Leckie

First exam published in 2001.

Published by Leckie & Leckie, 8 Whitehill Terrace, St. Andrews, Scotland KY16 8RN tel: 01334 475656 fax: 01334 477392 enquiries@leckieandleckie.co.uk www.leckieandleckie.co.uk

ISBN 1-84372-293-3

A CIP Catalogue record for this book is available from the British Library.

Printed in Scotland by Scotprint.

Leckie & Leckie is a division of Granada Learning Limited, part of ITV plc.

Acknowledgements

Leckie & Leckie is grateful to the copyright holders, as credited at the back of the book, for permission to use their material.

Every effort has been made to trace the copyright holders and to obtain their permission to use their copyright material.

Leckie & Leckie will gladly receive information enabling them to rectify any error or omission in subsequent editions.

[BLANK PAGE]

C

KU   PS

Total Marks

0500/402

NATIONAL
QUALIFICATIONS
2001

THURSDAY, 24 MAY
10.50 AM – 12.20 PM

CHEMISTRY
STANDARD GRADE
Credit Level

Fill in these boxes and read what is printed below.

Full name of centre

Town

Forename(s)

Surname

Date of birth
Day Month Year   Scottish candidate number   Number of seat

1 All questions should be attempted.

2 Necessary data will be found in the Data Booklet provided for Chemistry at Standard Grade and Intermediate 2.

3 The questions may be answered in any order but all answers are to be written in this answer book, and must be written clearly and legibly in ink.

4 Rough work, if any should be necessary, as well as the fair copy, is to be written in this book.

   Rough work should be scored through when the fair copy has been written.

5 Additional space for answers and rough work will be found at the end of the book.

6 The size of the space provided for an answer should not be taken as an indication of how much to write. It is not necessary to use all the space.

7 Before leaving the examination room you must give this book to the invigilator. If you do not, you may lose all the marks for this paper.

SCOTTISH
QUALIFICATIONS
AUTHORITY

PART 1

In **Questions 1 to 9** of this part of the paper, an answer is given by circling the appropriate letter (or letters) in the answer grid provided.

In some questions, **two** letters are required for full marks.

If more than the correct number of answers is given, marks will be deducted.

In some cases, the number of correct responses is **NOT** identified in the question.

A total of **20 marks** is available in this part of the paper.

SAMPLE QUESTION

A CH_4	B H_2	C CO_2
D CO	E C_2H_5OH	F C

(a) Identify the hydrocarbon(s).

(A)	B	C
D	E	F

The one correct answer to part (a) is A. This should be circled.

(b) Identify the **two** elements.

A	(B)	C
D	E	(F)

As indicated in this question, there are **two** correct answers to part (b). These are B and F. Both answers are circled.

(c) Identify the substance(s) which can burn to produce **both** carbon dioxide and water.

(A)	B	C
D	(E)	F

There are **two** correct answers to part (c). These are A and E.

Both answers are circled.

If, after you have recorded your answer, you decide that you have made an error and wish to make a change, you should cancel the original answer and circle the answer you now consider to be correct. Thus, in part (a), if you want to change an answer A to an answer D, your answer sheet would look like this:

(Ⱥ)	B	C
(D)	E	F

If you want to change back to an answer which has already been scored out, you should enter a tick (✓) in the box of the answer of your choice, thus:

✓(Ⱥ)	B	C
(Ð)	E	F

	KU	PS

1. Iron can be coated with different materials which provide a physical barrier against corrosion.

A	tin
B	grease
C	paint
D	plastic
E	zinc

(a) Identify the coating which also provides sacrificial protection.

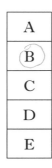

A
B
C
D
E

(b) Identify the coating which, if scratched, would cause the iron to rust faster than normal.

A
B
C
D
E

[Turn over

2. Frank and Dave carried out several experiments with metals and acids.

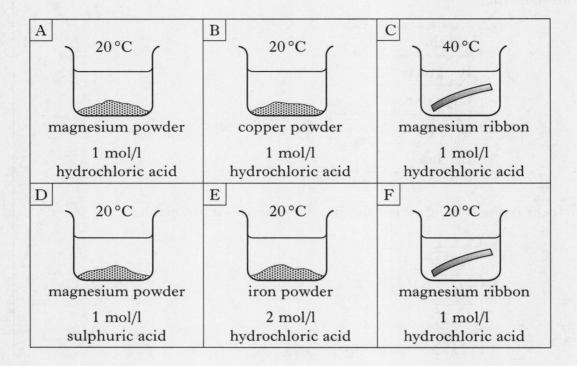

(a) Identify the **two** experiments which should be compared to show the effect of particle size on reaction rate.

A	B	C
D	E	F

(b) Identify the experiment in which no reaction would take place.

A	B	C
D	E	F

3.

A $^{24}_{11}Na$	B $^{14}_{6}C$	C $^{19}_{9}F$
D $^{24}_{12}Mg^{2+}$	E $^{19}_{9}F^{-}$	F $^{12}_{6}C$

(a) Identify the **two** particles with the same number of neutrons.

A	B	C
D	E	F

(b) Identify the **two** atoms which are isotopes.

A	B	C
D	E	F

(c) Identify the **two** particles with the same electron arrangement as neon.

A	B	C
D	E	F

[Turn over

4. The equations represent chemical reactions involving carbohydrates.

A	carbon dioxide + water → glucose + oxygen
B	glucose → starch + water
C	starch + water → glucose
D	glucose → ethanol + carbon dioxide
E	glucose + oxygen → carbon dioxide + water

(a) Identify the reaction which is catalysed by enzymes in yeast.

A
B
C
(D)
E

(b) Identify the hydrolysis reaction.

A
B
(C)
D
E

(c) Identify the reaction which takes place in animals during respiration.

A
B
C
D
(E)

5. The grid shows the names of some chemical compounds.

A sodium hydroxide	B potassium nitrate	C sodium chloride
D lithium carbonate	E sodium phosphate	F barium sulphate

(a) Identify the **two** bases.

A	B	C
D	E	F

(b) Identify the compound which could be prepared by precipitation.
You may wish to refer to page 5 of the data booklet.

A	B	C
D	E	F

[Turn over

6. The grid contains information about the particles found in atoms.

A	B	C
relative mass = 1	charge = 1+	found inside the nucleus
D	**E**	**F**
charge = 1–	relative mass almost zero	charge = zero

Identify the term(s) which can be applied to **both** protons **and** neutrons.

A	B	C
D	E	F

7. Several conductivity experiments were carried out using the apparatus shown below.

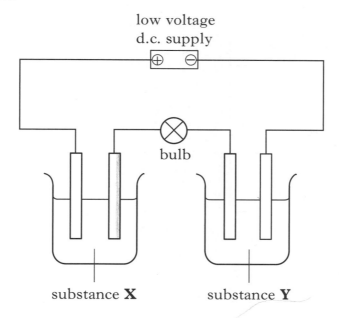

Identify the experiment(s) in which the bulb would light.

Experiment	Substance X	Substance Y
A	glucose solution	sodium chloride solution
B	molten tin	liquid mercury
C	sodium chloride solution	hexane
D	nickel bromide solution	molten sodium chloride
E	solid potassium nitrate	copper sulphate solution

A
B
C
D
E

[Turn over

Official SQA Past Papers: Credit Chemistry 2001

DO NOT
WRITE IN
THIS
MARGIN

KU | PS

8. The table below shows the names and colours of some common ions.

Ion	Formula	Colour
copper	Cu^{2+}	blue
nickel	Ni^{2+}	green
zinc	Zn^{2+}	colourless
lithium	Li^+	colourless
magnesium	Mg^{2+}	colourless
nitrate	NO_3^-	colourless
sulphate	SO_4^{2-}	colourless
permanganate	MnO_4^-	purple
dichromate	$Cr_2O_7^{2-}$	orange

Identify the true statement(s) based on the information in the table.

A	Copper nitrate is blue.
B	Coloured ions contain transition metals.
C	Ions containing oxygen are colourless.
D	All transition metal ions are coloured.
E	All lithium compounds are colourless.

A
B
C
D
E

	KU	PS

9. To turn a gas into a liquid it must be cooled below a temperature known as its critical temperature.

Gas	Formula	Relative formula mass	Critical temperature/°C
hydrogen	H_2	2	−240
helium	He	4	−268
ammonia	NH_3	17	133
oxygen	O_2	32	−119
carbon dioxide	CO_2	44	31

Identify the true statement(s) based on the information in the table.

A	Compounds have higher critical temperatures than elements.
B	Critical temperature increases as relative formula mass increases.
C	Diatomic elements have higher critical temperatures than monatomic elements.
D	Carbon dioxide can be a liquid at 40 °C.

A
B
C
D

[Turn over

DO NOT
WRITE IN
THIS
MARGIN

Marks | KU | PS

PART 2

A total of 40 marks is available in this part of the paper.

10. Andrew investigated the effect of different hydrocarbons on bromine solution.

Hydrocarbon	Formula	Effect on bromine solution
A	C_5H_{12}	
B	C_6H_{12}	no effect
C	C_5H_{10}	no effect
D	C_5H_{10}	quickly decolourised

(a) Complete the table to show the effect of hydrocarbon **A** on bromine solution.

1

(b) Name hydrocarbon **B**.

Hexene

1

(c) What term is used to describe a pair of hydrocarbons like **C** and **D**?

~~Isotopes~~ Alkenes

1

(3)

Marks KU PS

11. Siobhan carried out some experiments with four metals (**W**, **X**, **Y** and **Z**) and some of their compounds. She made the following observations.

When each metal was placed in cold water, only metal Y reacted.

Only metal W was obtained from its oxide by heating.

When metal X was placed in a solution containing ions of metal Z, metal X dissolved and solid metal Z was formed.

(*a*) Name the gas formed when metal **Y** reacts with water.

CO_2 1

(*b*) Suggest names for metals **W** and **Y**.

metal **W** _____ metal **Y** _____ 1

(*c*) Place the four metals (**W**, **X**, **Y** and **Z**) in order of reactivity (most reactive first).

_____ 1

(*d*) Name the type of chemical reaction which takes place when a metal is extracted from its oxide.

_____ 1

(4)

[Turn over

12. Some sources of methane contain hydrogen sulphide (H_2S).
This is removed before the methane is used as a fuel.

(a) Balance the equation for the combustion of methane.

$$CH_4 \quad + \quad O_2 \quad \rightarrow \quad CO_2 \quad + \quad H_2O$$

1

(b) Why is hydrogen sulphide removed before the methane is used as a fuel?

_____

_____

_____

1

(c) Hydrogen sulphide is removed by reacting it with sulphur dioxide.

$$2H_2S \quad + \quad SO_2 \quad \rightarrow \quad 2H_2O \quad + \quad 3S$$

Calculate the mass of sulphur produced, in grams, when 34 g of hydrogen sulphide reacts with sulphur dioxide.
Show your working clearly.

2

Marks | KU | PS

12. **(continued)**

(d) The table shows the relationship between solubility of sulphur dioxide in water and the temperature of the water.

Temperature/°C	0	10	20	30	40	60	80
Solubility/ grams per litre	225	145	95	60	35	15	5

(i) Draw a line graph of solubility against temperature.

Use appropriate scales to fill most of the graph paper.

(Additional graph paper, if required, will be found on page 24.)

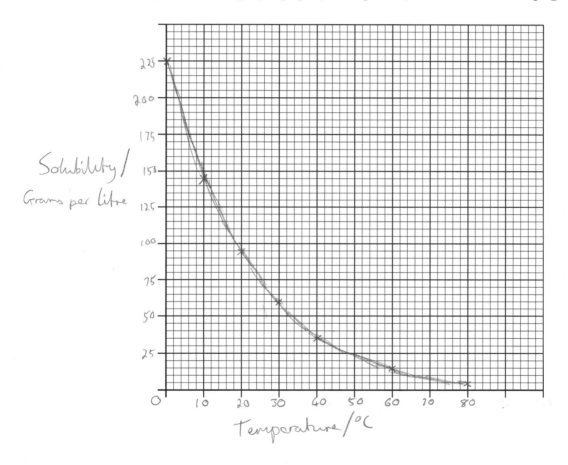

2

(ii) State the relationship between the solubility of sulphur dioxide in water and the temperature of the water.

_____

_____

1

(7)

[Turn over

Marks

KU | PS

13. Nitrogen forms many useful compounds.

Compound	Formula
Y	$(NH_4)_3PO_4$
potassium nitrate	KNO_3
urea	$CO(NH_2)_2$

(a) (i) Name compound **Y**.

_____   1

(ii) Compound **Y** can be used as a fertiliser.
Why are fertilisers added to the soil?

_____

_____

_____   1

(b) Which acid is used to make potassium nitrate?

_____   1

(c) Urea can be used to make a thermosetting polymer.

(i) What is meant by the term "thermosetting"?

_____

_____   1

(ii) Calculate the percentage mass of nitrogen in urea.
Show your working clearly.

2

(6)

Marks KU PS

14. The table compares the mass of ions found in ocean water with the mass of ions found in water from the Dead Sea.

Ion	Mass in 1 litre of ocean water/g	Mass in 1 litre of Dead Sea water/g
Na^+	10·7	31·5
K^+	0·4	6·8
Mg^{2+}	1·3	36·2
Ca^{2+}	0·4	13·4
Cl^-	19·2	183·0
Br^-	0·1	5·2
SO_4^{2-}	2·5	0·6

(a) What general statement can be made about the mass of ions in water from the Dead Sea compared with ocean water?

_____

_____

_____ 1

(b) Suggest a name for a compound which might be obtained if a sample of water from the Dead Sea was evaporated to dryness.

_____ 1

(c) Calculate the concentration of calcium ions, in mol/l, in ocean water.

1

(3)

[Turn over

15. Sarah set up the circuit shown below.

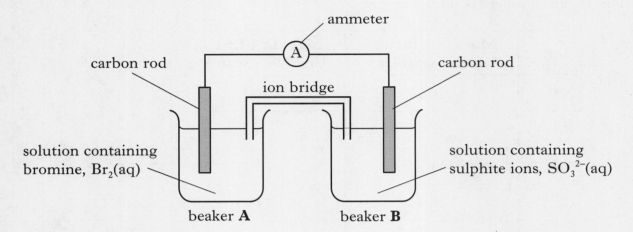

In beaker **B** sulphite ions are converted to sulphate ions:

$$SO_3^{2-}(aq) \quad + \quad H_2O(\ell) \quad \rightarrow \quad SO_4^{2-}(aq) \quad + \quad 2H^+(aq) \quad + \quad 2e^-$$

(a) On the diagram, clearly mark the path and the direction of the electron flow.

1

(b) (i) What term is used to describe the type of chemical reaction taking place in beaker **B**?

_____

1

(ii) Suggest what would happen to the pH in beaker **B**.

_____

1

(c) Write the ion-electron equation for the chemical reaction taking place in beaker **A**.

You may wish to use the data booklet to help you.

_____

1

(4)

DO NOT WRITE IN THIS MARGIN

Marks KU PS

16. (*a*) Ammonia is made industrially by the Haber process.

Name the catalyst used to make ammonia.

______________________________ **1**

(*b*) Name **two** compounds, which can react together to produce ammonia in the laboratory.

______________________________ **1**

(*c*) The atoms in an ammonia molecule are held together by covalent bonds. A covalent bond is a shared pair of electrons.

Explain how this holds the atoms together.

____

____

____

____

______________________________ **1**

(3)

[Turn over

Marks | KU | PS

17. (*a*) Methoxyethane belongs to a homologous series of compounds called ethers.

What is meant by the term "homologous series"?

_____

_____

_____   1

(*b*) Methoxyethane is formed when bromomethane, ethanol and sodium react together.

$2\ CH_3Br$ + $2\ C_2H_5OH$ + $2\ Na$ → $2\ CH_3OC_2H_5$ + $2\ NaBr$ + $\mathbf{X_2}$
bromomethane   ethanol   methoxyethane

(i) Name $\mathbf{X_2}$.

_____   1

(ii) Draw a **full** structural formula for methoxyethane ($CH_3OC_2H_5$).

1
(3)

Marks | KU | PS

18. Sea water contains magnesium ions. The diagram below shows how magnesium can be extracted from sea water.

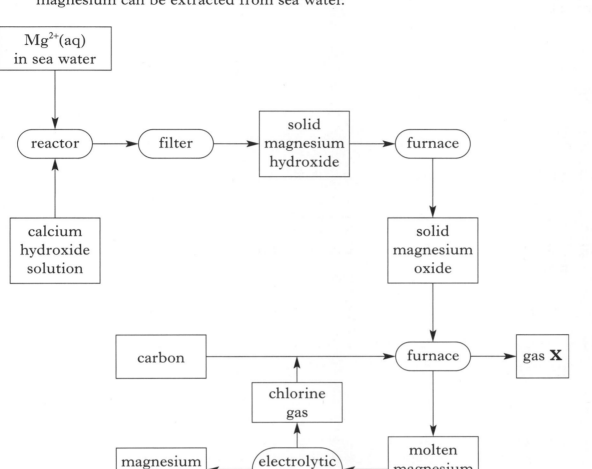

(a) Name the type of chemical reaction which takes place in the reactor.

_____ 1

(b) Write the **ionic** formula for calcium hydroxide.

_____ 1

(c) Name gas **X**.

_____ 1

(d) Why do ionic compounds like magnesium chloride conduct electricity when molten?

_____

_____

_____ 1

(4)

Marks | KU | PS

19. The alkynes are a family of hydrocarbons which contain a carbon to carbon triple bond.

$$H-C\equiv C-\overset{\displaystyle H}{\underset{\displaystyle H}{\overset{|}{\underset{|}{C}}}}-H \qquad\qquad H-C\equiv C-\overset{\displaystyle H}{\underset{\displaystyle H}{\overset{|}{\underset{|}{C}}}}-\overset{\displaystyle H}{\underset{\displaystyle H}{\overset{|}{\underset{|}{C}}}}-H$$

eg

propyne   butyne

(a) Suggest a general formula for the alkynes.

_____

1

(b) Alkynes are prepared by reacting a dibromoalkane with sodium hydroxide.

$$H-\overset{\displaystyle H}{\underset{\displaystyle Br}{\overset{|}{\underset{|}{C}}}}-\overset{\displaystyle H}{\underset{\displaystyle Br}{\overset{|}{\underset{|}{C}}}}-\overset{\displaystyle H}{\underset{\displaystyle H}{\overset{|}{\underset{|}{C}}}}-H \;+\; 2NaOH \;\rightarrow\; H-C\equiv C-\overset{\displaystyle H}{\underset{\displaystyle H}{\overset{|}{\underset{|}{C}}}}-H \;+\; 2NaBr \;+\; 2H_2O$$

dibromoalkane

(i) Draw the structural formula for the alkyne formed when the dibromoalkane shown below reacts with sodium hydroxide.

$$H-\overset{\displaystyle H}{\underset{\displaystyle H}{\overset{|}{\underset{|}{C}}}}-\overset{\displaystyle H}{\underset{\displaystyle Br}{\overset{|}{\underset{|}{C}}}}-\overset{\displaystyle H}{\underset{\displaystyle Br}{\overset{|}{\underset{|}{C}}}}-\overset{\displaystyle H}{\underset{\displaystyle H}{\overset{|}{\underset{|}{C}}}}-\overset{\displaystyle H}{\underset{\displaystyle H}{\overset{|}{\underset{|}{C}}}}-H$$

$\downarrow$

1

DO NOT WRITE IN THIS MARGIN

Marks | KU | PS

19. (*b*) (continued)

(ii) Suggest why the dibromoalkane shown below does **not** form an alkyne when heated with sodium hydroxide.

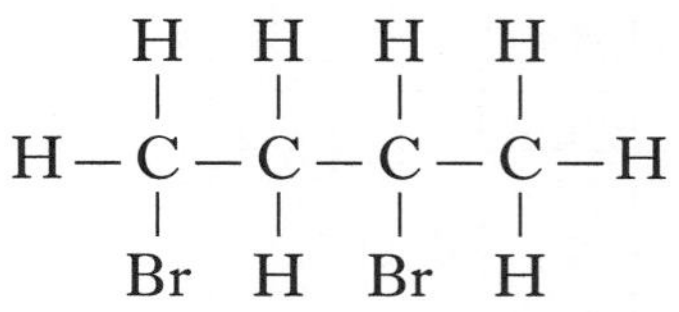
H H H H
H–C–C–C–C–H
Br H Br H

____

____

____ **1**

(3)

[*END OF QUESTION PAPER*]

ADDITIONAL SPACE FOR ANSWERS

ADDITIONAL GRAPH PAPER FOR QUESTION 12(*d*)(i)

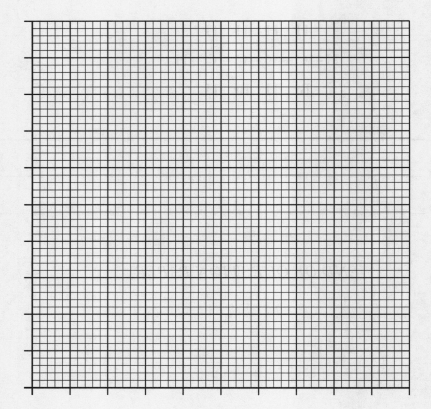

[BLANK PAGE]

FOR OFFICIAL USE

KU   PS

Total Marks

0500/402

NATIONAL
QUALIFICATIONS
2002

THURSDAY, 16 MAY
2.50 PM – 4.20 PM

CHEMISTRY
STANDARD GRADE
Credit Level

Fill in these boxes and read what is printed below.

Full name of centre

Town

Forename(s)

Surname

Date of birth
Day Month Year

Scottish candidate number

Number of seat

1 All questions should be attempted.

2 Necessary data will be found in the Data Booklet provided for Chemistry at Standard Grade and Intermediate 2.

3 The questions may be answered in any order but all answers are to be written in this answer book, and must be written clearly and legibly in ink.

4 Rough work, if any should be necessary, as well as the fair copy, is to be written in this book.

   Rough work should be scored through when the fair copy has been written.

5 Additional space for answers and rough work will be found at the end of the book.

6 The size of the space provided for an answer should not be taken as an indication of how much to write. It is not necessary to use all the space.

7 Before leaving the examination room you must give this book to the invigilator. If you do not, you may lose all the marks for this paper.

SCOTTISH
QUALIFICATIONS
AUTHORITY

PART 1

In Questions 1 to 9 of this part of the paper, an answer is given by circling the appropriate letter (or letters) in the answer grid provided.

In some questions, two letters are required for full marks.

If more than the correct number of answers is given, marks will be deducted.

In some cases, the number of correct responses is NOT identified in the question.

A total of 20 marks is available in this part of the paper.

SAMPLE QUESTION

A CH_4	B H_2	C CO_2
D CO	E C_2H_5OH	F C

(a) Identify the hydrocarbon(s).

Ⓐ	B	C
D	E	F

The one correct answer to part (a) is A. This should be circled.

(b) Identify the **two** elements.

A	Ⓑ	C
D	E	Ⓕ

As indicated in this question, there are **two** correct answers to part (b). These are B and F. Both answers are circled.

(c) Identify the substance(s) which can burn to produce **both** carbon dioxide and water.

Ⓐ	B	C
D	Ⓔ	F

There are **two** correct answers to part (c). These are A and E.

Both answers are circled.

If, after you have recorded your answer, you decide that you have made an error and wish to make a change, you should cancel the original answer and circle the answer you now consider to be correct. Thus, in part (a), if you want to change an answer A to an answer D, your answer sheet would look like this:

Ⓐ̶	B	C
Ⓓ	E	F

If you want to change back to an answer which has already been scored out, you should enter a tick (✓) in the box of the answer of your choice, thus:

✓Ⓐ̶	B	C
Ⓓ̶	E	F

1. The grid shows the names of some common ionic compounds.

A	B	C
ammonium chloride	calcium carbonate	potassium chloride
D	E	F
calcium sulphate	magnesium sulphate	sodium carbonate

(a) Identify the **two** compounds which could be used as fertilisers.

A	B	C
D	E	F

(b) Identify the **two** compounds which are bases.

A	B	C
D	E	F

[Turn over

2.

Substance	Conducts as		Melting point/°C
	a solid	a liquid	
A	no	yes	801
B	no	no	113
C	yes	yes	63
D	no	no	1700
E	yes	yes	98
F	no	no	44

(a) Identify the substance which could be sodium chloride.

A
B
C
D
E
F

(b) Identify the **two** substances which exist as molecules.

A
B
C
D
E
F

DO NOT WRITE IN THIS MARGIN

KU | PS

3. The symbols for some elements are shown below.

A Li	B O	C Mg
D Si	E F	F K

(*a*) Identify the **two** elements which form an ionic compound with a formula of the type XY_2, where X is a metal.

A	B	C
D	E	F

(*b*) Identify the **two** elements which would react together to form molecules with the same shape as a methane molecule.

A	B	C
D	E	F

[Turn over

Official SQA Past Papers: Credit Chemistry 2002

DO NOT
WRITE IN
THIS
MARGIN

KU PS

4. Hydrocarbons contain hydrogen and carbon only.

A	B	C
H H H H | | | | H—C—C—C—C—H | | | | H H H H	H H | | C=C — C—H | | | H CH_3 H	H H | | H—C—C—H | | H—C—C—H | | H H
D	E	F

H   H   H   H
| | | |
C=C—C—C—H
| | |
H   H   H
	H H
	\ H—C—C—H /
	         C
	      / \
	   H H
		H H H
		| | |
		H—C—C—C—H
		| | |
		H CH_3 H

(a) Identify the **two** hydrocarbons which would quickly decolourise bromine solution.

A	B	C
D	E	F

(b) Identify the isomer of the hydrocarbon in box D which belongs to a different homologous series.

A	B	C
D	E	F

5.

Particle	Number of		
	protons	neutrons	electrons
A	12	13	12
B	8	10	10
C	12	12	10
D	10	12	10
E	8	10	8

(a) Identify the particle which is a positive ion.

A
B
C
D
E

(b) Identify the **two** particles which are isotopes.

A
B
C
D
E

[Turn over

Page seven

6. An atom of carbon can be represented by the symbol $^{14}_{6}C$.

Identify the correct statement(s) about this carbon atom.

A	It has 14 protons.
B	It has 8 neutrons.
C	It has more protons than neutrons.
D	It has an equal number of protons and neutrons.
E	It has an equal number of protons and electrons.
F	It has an equal number of neutrons and electrons.

A
B
C
D
E
F

7.

A	B
C_4H_{10} + O_2	$CaCO_3$ + HCl
C	D
Zn + H_2SO_4	Li + H_2O
E	F
CuO + C	Cu + $ZnSO_4$

(a) Which box contains a pair of chemicals that will **not** react with each other?

A	B
C	D
E	F

(b) Which box(es) contain(s) a pair of chemicals that react to form water?

A	B
C	D
E	F

[Turn over

Official SQA Past Papers: Credit Chemistry 2002

DO NOT
WRITE IN
THIS
MARGIN

KU | PS

8. Equations are used to represent chemical reactions.

A	$2H_2(g) + O_2(g) \rightarrow 2H_2O(g)$
B	$Zn(s) + FeSO_4(aq) \rightarrow Fe(s) + ZnSO_4(aq)$
C	$Fe^{2+}(aq) \rightarrow Fe^{3+}(aq) + e^-$
D	$CH_4(g) + 2O_2(g) \rightarrow CO_2(g) + 2H_2O(g)$
E	$2H_2O(\ell) + O_2(g) + 4e^- \rightarrow 4OH^-(aq)$
F	$Fe^{2+}(aq) + 2e^- \rightarrow Fe(s)$

(a) Identify the **two** equations which represent combustion reactions.

A
B
C
D
E
F

(b) Identify the equation(s) which represent(s) a step in the rusting of iron.

A
B
C
D
E
F

DO NOT WRITE IN THIS MARGIN

KU | PS

9. Identify the statement(s) which refer(s) to an atom of fluorine.

You may wish to use the data booklet to help you.

A	It has a stable electron arrangement.
B	It will form an ion by losing one electron.
C	It will form an ion with a single negative charge.
D	It has two more electrons than an oxygen atom.
E	It has the same number of electrons as a chlorine atom.
F	It has the same number of outer electrons as an iodine atom.

A
B
C
D
E
F

[Turn over

DO NOT WRITE IN THIS MARGIN

Marks | KU | PS

PART 2

A total of 40 marks is available in this part of the paper.

10. Ethene is a starting material in the manufacture of the polymer poly(vinylchloride), PVC.

(a) Name the process used to make ethene from hydrocarbons obtained from crude oil.

_____ 1

(b) Part of a PVC molecule is shown below.

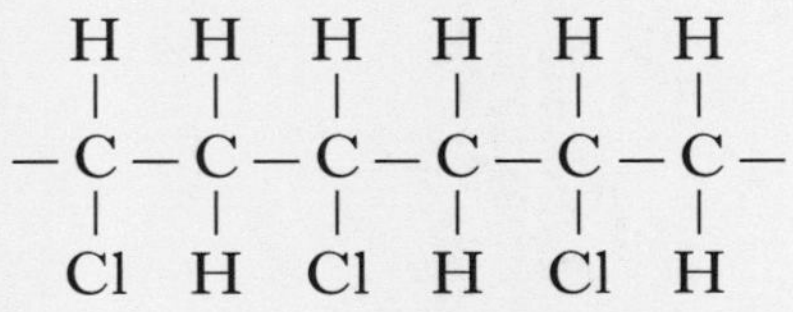

$$
\begin{array}{cccccc}
\text{H} & \text{H} & \text{H} & \text{H} & \text{H} & \text{H} \\
| & | & | & | & | & | \\
-\text{C} - \text{C} - \text{C} - \text{C} - \text{C} - \text{C}- \\
| & | & | & | & | & | \\
\text{Cl} & \text{H} & \text{Cl} & \text{H} & \text{Cl} & \text{H}
\end{array}
$$

(i) Draw the structure of the repeating unit in a PVC molecule.

1

(ii) Name a toxic gas produced when PVC burns.

_____ 1

(3)

Marks | KU | PS

11. (*a*) Ailsa carried out the experiment shown below.

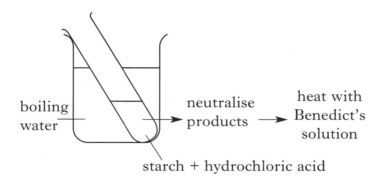

boiling water → starch + hydrochloric acid → neutralise products → heat with Benedict's solution

**Result:
Benedict's solution
turns red/orange**

(i) What type of chemical reaction takes place when starch is heated with hydrochloric acid?

_____ 1

(ii) Ailsa said that the starch had turned into glucose.
Name another sugar which turns Benedict's solution red/orange.

_____ 1

(iii) Ailsa repeated her experiment using amylase solution instead of hydrochloric acid.
Suggest a reason why the Benedict's solution did not turn red/orange.

_____ 1

(*b*) Write the molecular formula for glucose.

_____ 1

(4)

[**Turn over**

Marks | KU | PS

12. Titanium compounds have many uses.

(a) (i) Warships can produce a smokescreen by reacting titanium(IV) chloride with water:

$$TiCl_4(\ell) \quad + \quad H_2O(\ell) \quad \rightarrow \quad TiO_2(s) \quad + \quad HCl(g)$$

Balance this equation. **1**

(ii) Titanium(IV) chloride is a liquid at room temperature.

What type of bonding does this suggest is present in titanium(IV) chloride?

_____ **1**

(b) Titanium(IV) oxide (TiO_2) is used as a white pigment in paint.

Calculate the percentage by mass of titanium in TiO_2.

(Relative atomic mass of titanium = 48)

Show your working clearly.

2

(4)

Marks | KU | PS

13. Copper displaces silver from silver(I) nitrate solution.

$$Cu(s) + 2Ag^+(aq) + 2NO_3^-(aq) \rightarrow Cu^{2+}(aq) + 2NO_3^-(aq) + 2Ag(s)$$

(a) Rewrite the equation omitting the spectator ions.

1

(b) Write the ion-electron equation for the oxidation step in the displacement reaction.

You may wish to use the data booklet to help you.

1

(c) The reaction can also be carried out in a cell.

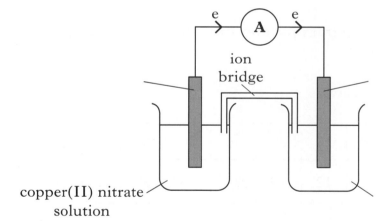

copper(II) nitrate
solution

(i) Complete the three labels on the diagram.

1

(ii) The purpose of the ion bridge is to complete the circuit.

Suggest why sodium carbonate solution should not be used in the ion bridge.

You may wish to use the data booklet to help you.

_____

_____

_____ 1

(4)

Marks | KU | PS

14. Sodium carbonate reacts with hydrochloric acid to form carbon dioxide. Brian measured the volume of carbon dioxide given off over a period of time and recorded his results.

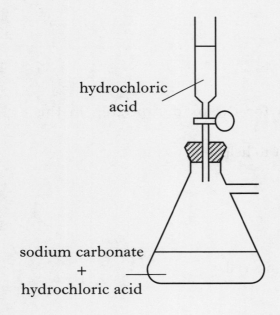

hydrochloric
acid

sodium carbonate
+
hydrochloric acid

(a) Complete and label the diagram to show how Brian measured the volume of carbon dioxide.

2

(b) Brian's results are shown below.

Time/s	0	10	30	40	50	60	70
Volume of carbon dioxide/cm³	0	12	29	34	36	37	37

14. *(b)* **(continued)**

Draw a line graph of the results.

Use appropriate scales to fill most of the graph paper.

(Additional graph paper, if required, will be found on page 26.)

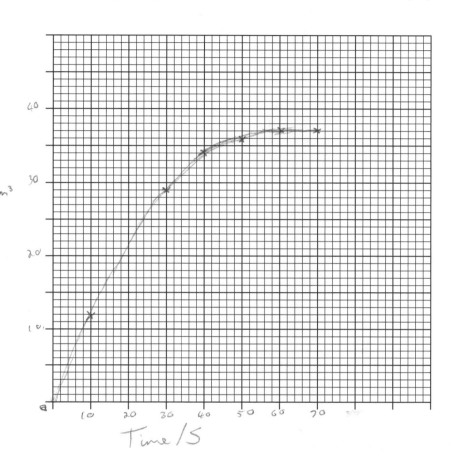

Volume of CO_2/cm^3

Time /S

2

(c) Suggest a value for the volume of carbon dioxide collected during the first 20 seconds.

_____ cm^3

1

(d) Write the ionic formula for sodium carbonate.

_____

1

(6)

[Turn over

Marks | KU | PS

15. Some metals are found uncombined in the Earth's crust but others have to be extracted from their ores.

(a) Place the following metals in the correct space in the table.

lead, magnesium, mercury

You may wish to use the data booklet to help you.

Metal	Method of Extraction
	using heat alone
	using heat and carbon
	electrolysis of molten compound

1

(b) Iron is extracted by reacting iron(III) oxide with carbon monoxide.

(i) Name the type of industrial plant where iron is extracted.

_____

1

(ii) The overall reaction taking place during the extraction of iron is given by the equation:

$$Fe_2O_3 \ + \ 3CO \ \rightarrow \ 2Fe \ + \ 3CO_2$$

Calculate the mass of iron, in tonnes, which is produced from 1600 tonnes of iron(III) oxide.

Show your working clearly.

Answer: _____ tonnes

2

(4)

Marks | KU | PS

16. Fermentation is used to produce alcohol from sugars like glucose.

(a) Name the gas produced during the fermentation of glucose.

_____

1

(b) Why does fermentation stop when the alcohol concentration reaches approximately 15 %?

_____

_____

1

(c) In industry, ethanol (alcohol) can be produced from ethene as shown below.

$$H-\underset{\underset{H}{|}}{\overset{\overset{H}{|}}{C}}=\underset{\underset{H}{|}}{\overset{\overset{}{}}{C}}-H \quad + \quad H_2O \quad \longrightarrow \quad H-\underset{\underset{H}{|}}{\overset{\overset{H}{|}}{C}}-\underset{\underset{H}{|}}{\overset{\overset{OH}{|}}{C}}-H$$

ethene   ethanol

(i) Name the type of chemical reaction taking place.

_____

1

(ii) Draw a structural formula for the product of the following reaction:

$$H-\overset{\overset{}{}}{C}=\underset{\underset{CH_3}{|}}{\overset{\overset{}{}}{C}}-\underset{\underset{H}{|}}{\overset{\overset{H}{|}}{C}}-\underset{\underset{H}{|}}{\overset{\overset{H}{|}}{C}}-H \quad + \quad H_2O$$

$$\downarrow$$

1

(4)

[Turn over

Marks | KU | PS

17. Alcohols can be oxidised by hot copper(II) oxide.

The product is either an aldehyde or a ketone.

Alcohol	Structural formula	Type of product	Structural formula
ethanol	H H | | H—C—C—OH | | H H	an aldehyde	H O | // H—C—C | \ H H
propan-1-ol	H H H | | | H—C—C—C—OH | | | H H H	an aldehyde	H H O | | // H—C—C—C | | \ H H H
propan-2-ol	H H H | | | H—C—C—C—H | | | H OH H	a ketone	H H | | H—C—C—C—H | || | H O H
butan-2-ol	H H H H | | | | H—C—C—C—C—H | | | | H OH H H	a ketone	H H H | | | H—C—C—C—C—H | || | | H O H H

(*a*) (i) Aldehydes and ketones have the same general formula.

Suggest a general formula for these compounds.

_____   1

(ii) Write a general statement linking the type of product to the structure of the alcohol used.

_____

_____

_____   1

DO NOT WRITE IN THIS MARGIN

Marks | KU | PS

17. (continued)

(*b*) In these reactions the copper(II) oxide is reduced to copper metal.
Suggest why aluminium oxide cannot be used to oxidise alcohols.

____

____

____ **1**

(3)

[Turn over

DO NOT
WRITE IN
THIS
MARGIN

Marks | KU | PS

18. The flow chart shows some processes which take place in an industrial chemical complex.

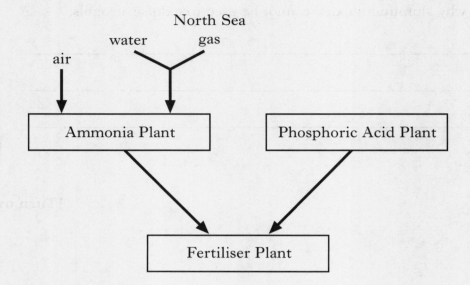

(*a*) Air and water are used as raw materials because they contain the elements needed to make ammonia.

Suggest **one** other reason why they are used as raw materials.

_____

_____ 1

(*b*) Which reactant for the ammonia plant must be produced in the reaction between North Sea gas and water?

_____ 1

(*c*) Name the salt formed in the fertiliser plant.

_____ 1

Official SQA Past Papers: Credit Chemistry 2002

DO NOT
WRITE IN
THIS
MARGIN

Marks KU PS

18. (continued)

(*d*) The graph shows the different percentage yields of ammonia which can be obtained under different conditions in the ammonia plant.

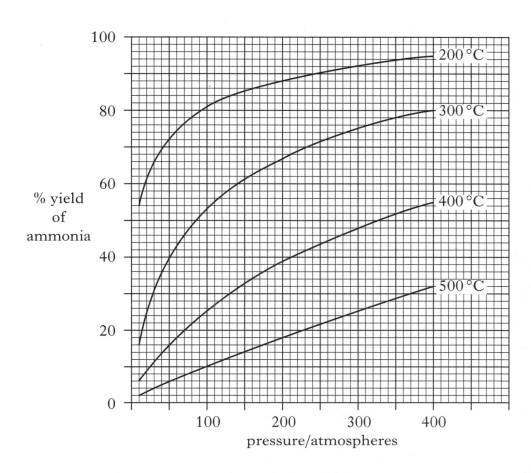

   (i) What is the relationship between the percentage yield of ammonia and the temperature at constant pressure?

   _____

   _____

   _____ **1**

   (ii) Explain why all of the nitrogen and hydrogen are not converted to ammonia.

   _____

   _____

   _____ **1**

   (5)

Marks KU PS

19. Vinegar is a dilute solution of ethanoic acid in water.

Karen carried out a titration to find out the concentration of ethanoic acid in some vinegar.

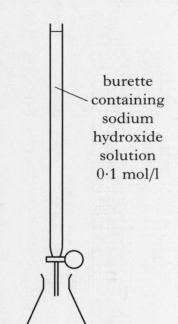

burette
containing
sodium
hydroxide
solution
0·1 mol/l

25 cm³ vinegar
plus indicator

	Rough titre	**1st titre**	**2nd titre**
Initial burette reading/cm³	1·0	21·7	11·7
Final burette reading/cm³	21·7	41·7	31·9
Volume used/cm³	20·7	20·0	20·2

(a) Karen used data from the table to calculate an average volume of sodium hydroxide solution.

She used this average volume to calculate the number of moles of sodium hydroxide needed to neutralise the acid in 25 cm³ of the vinegar.

(i) What average volume of sodium hydroxide should she have used?

_____ cm³

1

DO NOT WRITE IN THIS MARGIN

Marks | KU | PS

19. **(a)** **(continued)**

   (ii) Calculate the number of moles of sodium hydroxide in this average volume.

   Show your working clearly.

1

   (b) 1 mole of ethanoic acid reacts with 1 mole of sodium hydroxide.

   Calculate the concentration, in mol/l, of ethanoic acid in the vinegar.

   Show your working clearly

1

(3)

[END OF QUESTION PAPER]

ADDITIONAL SPACE FOR ANSWERS

ADDITIONAL GRAPH PAPER FOR QUESTION 14(*b*)

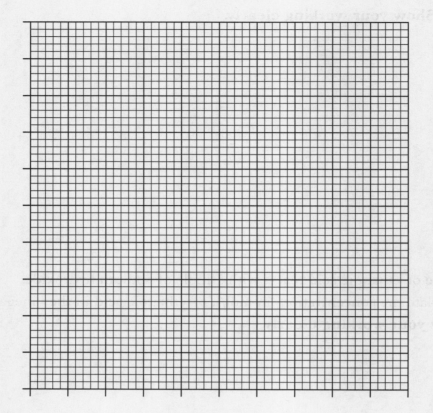

2003 | Credit

[BLANK PAGE]

FOR OFFICIAL USE

C

KU   PS

Total Marks

0500/402

NATIONAL
QUALIFICATIONS
2003

FRIDAY, 23 MAY
10.50 AM – 12.20 PM

CHEMISTRY
STANDARD GRADE
Credit Level

Fill in these boxes and read what is printed below.

Full name of centre

Town

Forename(s)

Surname

Date of birth
Day Month Year

Scottish candidate number

Number of seat

1 All questions should be attempted.

2 Necessary data will be found in the Data Booklet provided for Chemistry at Standard Grade and Intermediate 2.

3 The questions may be answered in any order but all answers are to be written in this answer book, and must be written clearly and legibly in ink.

4 Rough work, if any should be necessary, as well as the fair copy, is to be written in this book.

Rough work should be scored through when the fair copy has been written.

5 Additional space for answers and rough work will be found at the end of the book.

6 The size of the space provided for an answer should not be taken as an indication of how much to write. It is not necessary to use all the space.

7 Before leaving the examination room you must give this book to the invigilator. If you do not, you may lose all the marks for this paper.

SCOTTISH
QUALIFICATIONS
AUTHORITY

©

PART 1

In Questions 1 to 9 of this part of the paper, an answer is given by circling the appropriate letter (or letters) in the answer grid provided.

In some questions, two letters are required for full marks.

If more than the correct number of answers is given, marks will be deducted.

In some cases, the number of correct responses is NOT identified in the question.

A total of 20 marks is available in this part of the paper.

SAMPLE QUESTION

A CH_4	B H_2	C CO_2
D CO	E C_2H_5OH	F C

(a) Identify the hydrocarbon(s).

Ⓐ	B	C
D	E	F

The one correct answer to part (a) is A. This should be circled.

(b) Identify the **two** elements.

A	Ⓑ	C
D	E	Ⓕ

As indicated in this question, there are **two** correct answers to part (b). These are B and F. Both answers are circled.

(c) Identify the substance(s) which can burn to produce **both** carbon dioxide and water.

Ⓐ	B	C
D	Ⓔ	F

There are **two** correct answers to part (c). These are A and E.

Both answers are circled.

If, after you have recorded your answer, you decide that you have made an error and wish to make a change, you should cancel the original answer and circle the answer you now consider to be correct. Thus, in part (a), if you want to change an answer A to an answer D, your answer sheet would look like this:

(A̶)	B	C
Ⓓ	E	F

If you want to change back to an answer which has already been scored out, you should enter a tick (✓) in the box of the answer of your choice, thus:

✓(A̶)	B	C
(D̶)	E	F

1. Atoms are made up of protons, neutrons and electrons.

A	The number of protons
B	The number of neutrons
C	The number of electrons
D	The number of outer electrons
E	The number of protons plus neutrons

(a) Identify the **two** numbers which are the same in a neutral atom.

A
B
C
D
E

(b) Identify the mass number of an atom.

A
B
C
D
E

[Turn over

KU | PS

2. The names of several compounds are shown in the grid.

A potassium nitrate	B sodium hydroxide	C lithium sulphate
D aluminium chloride	E ammonium phosphate	F calcium chloride

(a) Identify the **two** compounds which can be used as fertilisers.

A	B	C
D	E	F

(b) Identify the **two** compounds which react together to produce ammonia.

A	B	C
D	E	F

Official SQA Past Papers: Credit Chemistry 2003

DO NOT
WRITE IN
THIS
MARGIN

KU | PS

3. Hydrocarbons are compounds made from hydrogen and carbon only.

(a) Identify the hydrocarbon which reacts with hydrogen to form butane.

A	B	C
D	E	F

(b) Identify the **two** isomers.

A	B	C
D	E	F

(c) Identify the hydrocarbon(s) which is (are) the first member(s) of a homologous series.

A	B	C
D	E	F

[Turn over

DO NOT
WRITE IN
THIS
MARGIN

KU | PS

4. The grid shows some pairs of chemicals.

A	sodium + water	B	zinc + magnesium sulphate solution
C	copper carbonate + dilute sulphuric acid	D	lead nitrate solution + potassium iodide solution
E	silver + dilute hydrochloric acid	F	potassium hydroxide solution + dilute nitric acid

Which box(es) contain(s) a pair of chemicals that react to form a gas?

A	B
C	D
E	F

5. Identify the result(s) obtained in the reaction between dilute sulphuric acid and barium hydroxide solution.

You may wish to use the data booklet to help you.

A	The pH of the acid went down.
B	Carbon dioxide was produced.
C	A precipitate was formed.
D	Hydrogen was produced.
E	Water was produced.

A
B
C
D
E

[Turn over

6. Oil rigs made from iron should be protected from rusting.
   Identify the correct statement(s).

A	Salt water slows down rusting.
B	Tin gives sacrificial protection to the iron.
C	The rusting of iron is an example of oxidation.
D	Ferroxyl indicator turns blue in the presence of Fe^{2+} ions.
E	Iron rusts faster when connected to the negative terminal of a battery.

A
B
C
D
E

7. Iron(III) oxide is an ionic compound.

Identify the correct statement(s).

A	It is a salt.
B	It can be reduced to iron.
C	It has the formula Fe_2O_3.
D	It is made up of molecules.
E	It does not react with acid.

A
B
C
D
E

[Turn over

8. The table contains information about some solid, liquid and gaseous compounds.

Compound	Melting point /°C	Boiling point /°C	pH of solution in water
A	319	1390	11
B	801	1413	7
C	−115	−85	3
D	−93	−6	11
E	−95	56	7
F	63	189	3

(*a*) Identify the compound which is a gas at 25 °C and forms an acidic solution.

A
B
C
D
E
F

(*b*) Identify the compound which could be sodium hydroxide.

A
B
C
D
E
F

	KU	PS

9. Substances can be classified as conductors or non-conductors and also as solids or liquids.

Substance	State	Conductor or non-conductor
A	solid	non-conductor
B	liquid	non-conductor
C	solid	conductor
D	liquid	conductor

(a) Which **two** substances could be sodium chloride?

A
B
C
D

(b) Which substance could **not** be a compound?

A
B
C
D

[Turn over

Marks | KU | PS

PART 2

A total of 40 marks is available in this part of the paper.

10. The diagrams show how different flames can be produced in a Bunsen burner.

Flame **A**   Flame **B**

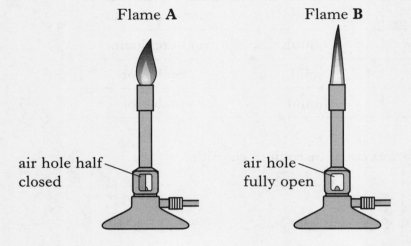

air hole half
closed

air hole
fully open

(a) The fuel used in a Bunsen burner is methane, CH_4.
What is meant by the term "fuel"?

_____

_____   **1**

(b) Methane burns to form carbon dioxide and water.

(i) Balance this equation.

$$CH_4 \quad + \quad O_2 \quad \rightarrow \quad CO_2 \quad + \quad H_2O$$

1

(ii) Name another product which could be formed in Flame **A**.

_____   **1**

(c) Draw a diagram to show the **shape** of a methane molecule.

1
(4)

11. There are two different types of chlorine atom: $^{35}_{17}Cl$ and $^{37}_{17}Cl$.

(a) (i) What name is used to describe these different types of chlorine atom?

_____ **1**

(ii) A natural sample of chlorine has an average atomic mass of 35·5. What is the mass number of the more abundant type of chlorine atom in the sample?

_____ **1**

(b) The atoms in a chlorine molecule are held together by a covalent bond. A covalent bond is a shared pair of electrons.
Explain how this holds the atoms together.

_____

_____

_____

_____

_____ **1**

(c) Complete the table to show the number of each type of particle in a $^{35}_{17}Cl^-$ ion.

Particle	Number
proton	
neutron	
electron	

2

(5)

[Turn over

Marks | KU | PS

12. Hydrogen can be produced in the laboratory by adding excess hydrochloric acid to lumps of zinc. The reaction stops when all the zinc is used up.

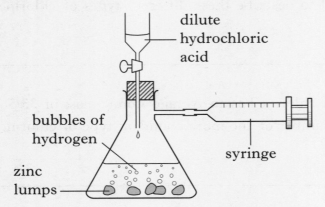

dilute hydrochloric acid

bubbles of hydrogen

zinc lumps

syringe

The volume of hydrogen gas produced over a period of time was measured and the results are shown in the table.

Time/s	0	20	40	60	80	100	120	140
Volume of hydrogen/cm^3	0	30	51	65	74	78	80	80

(*a*) Draw a line graph of the results.

Use appropriate scales to fill most of the graph paper.

(Additional graph paper, if required, will be found on page 24.)

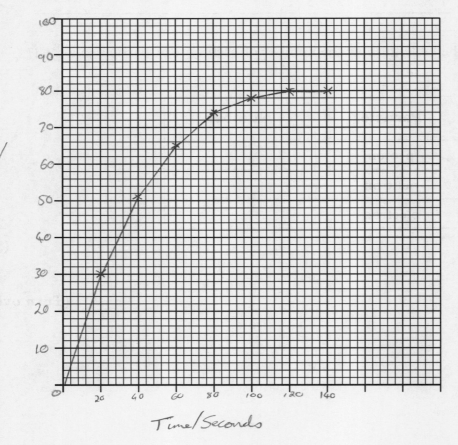

Vol of H$_2$/ cm^3

Time/Seconds

2

Page fourteen

DO NOT WRITE IN THIS MARGIN

Marks KU PS

12. (continued)

(*b*) Use your graph to estimate the time, in seconds, for 40 cm^3 of hydrogen to be produced.

____ **1**

(*c*) The equation for the reaction of zinc with hydrochloric acid is

$$Zn + 2HCl \rightarrow ZnCl_2 + H_2$$

Calculate the mass of zinc required to produce 0·5 mole of hydrogen.

Answer: ______________ g **1**

(4)

[Turn over

Marks KU PS

13. Mrs Smith gave her class three chemicals labelled **P**, **Q** and **R**.

The chemicals were ethanol (C_2H_5OH) solution, silver nitrate solution and dilute sulphuric acid.

The class used the following apparatus to identify each solution.

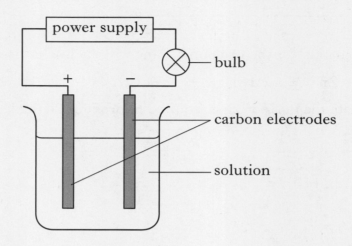

The results are shown in the table.

Solution	Bulb lights	Observation at electrodes
P	Yes	grey solid formed at negative electrode
Q	No	no reaction
R	Yes	a gas formed at both electrodes

(a) Identify **P**.

_____

1

(b) What type of bonding is present in **Q**?

_____

1

(c) Name the gas formed at the negative electrode when solution **R** is used.

_____

1

(d) What process would be used to obtain a sample of ethanol from the ethanol solution?

_____

1

(4)

Marks   KU   PS

14. Propene has the structural formula shown below.

$$\begin{array}{c} \text{H} \\ | \\ \overset{\text{H}}{\underset{\text{H}}{>}}\text{C} = \text{C} - \text{C} - \text{H} \\ \quad\quad | \quad | \\ \quad\quad \text{H} \quad \text{H} \end{array}$$

Propene quickly decolourises bromine water, $Br_2(aq)$.

(*a*)   (i)   Name the type of chemical reaction which takes place when propene reacts with bromine water.

_____   1

      (ii)   Draw the **full** structural formula for the product of the reaction.

1

(*b*)   Propene can be converted into the polymer, poly(propene).

Complete the diagram to show how **three** propene molecules join to form part of the polymer chain.

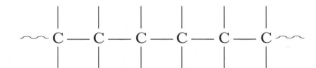

1

(3)

[Turn over

Marks

KU | PS

15. Bones are formed when calcium ions and phosphate ions combine to form insoluble calcium phosphate, $Ca_3(PO_4)_2$.

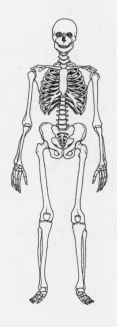

This reaction can be reproduced in the laboratory by adding a solution of calcium chloride to a solution of sodium phosphate.

$$3Ca^{2+}(aq) + 6Cl^-(aq) + 6Na^+(aq) + 2PO_4^{3-}(aq) \rightarrow 6Na^+(aq) + 6Cl^-(aq) + (Ca^{2+})_3(PO_4^{3-})_2(s)$$

(a) Circle the spectator ions in the above equation.

1

(b) What technique could be used to remove the calcium phosphate from the mixture?

_____

1

(c) Calculate the percentage by mass of calcium in calcium phosphate, $Ca_3(PO_4)_2$.

Answer: _____ %

2

(4)

Marks | KU | PS

16. Helen set up the cell shown below.

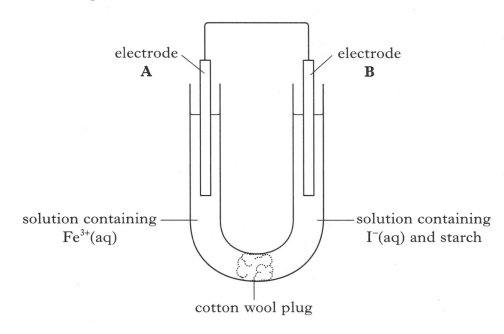

electrode **A**

electrode **B**

solution containing $Fe^{3+}(aq)$

solution containing $I^-(aq)$ and starch

cotton wool plug

The reaction taking place at electrode **A** is

$$Fe^{3+}(aq) \; + \; e^- \; \rightarrow \; Fe^{2+}(aq)$$

(a) (i) **On the diagram**, clearly mark the path and the direction of electron flow.

_____ 1

(ii) What term is used to describe the type of chemical reaction taking place at electrode **A**?

_____ 1

(b) Iodine forms at electrode **B**.

(i) What would you **see** happening around electrode **B**?

_____

_____ 1

(ii) Write an ion-electron equation for the chemical reaction taking place at electrode **B**.
You may wish to use the data booklet to help you.

_____ 1

(4)

[Turn over

Official SQA Past Papers: Credit Chemistry 2003

DO NOT
WRITE IN
THIS
MARGIN

Marks | KU | PS

17. The flow diagram shows what happens to starchy foods after they have been eaten.

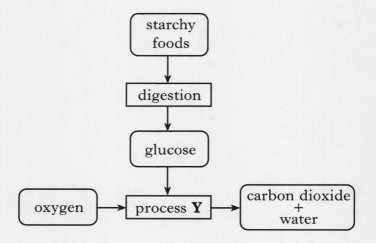

(a) What **type** of substance, present in the digestive system, speeds up the breakdown of starchy foods?

_____ 1

(b) What **type** of chemical reaction takes place when starch is broken down into glucose during digestion?

_____ 1

(c) Process **Y** provides the body with energy.
Name this process.

_____ 1

(d) Name an isomer of glucose.

_____ 1

                                                     (4)

Marks | KU | PS

18. The energy required to remove an outer electron from an atom is called the ionisation energy.

(a) The equation for the ionisation of a magnesium atom is

$$Mg(g) \longrightarrow Mg^+(g) + e^-$$

Write the electron arrangement for $Mg^+(g)$.

_____

1

(b) The graph shows the ionisation energy values for the first 20 elements.

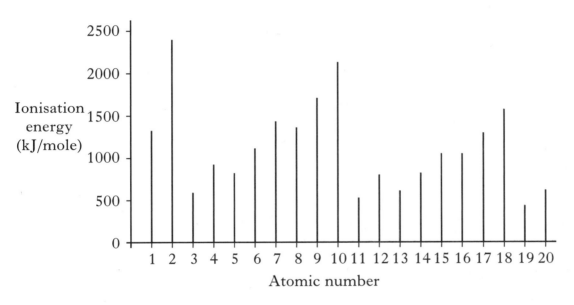

(i) Describe the general trend in ionisation energy going from lithium to neon.

_____

_____

_____

1

(ii) Describe the trend in ionisation energy going down a group.

_____

_____

_____

1

(3)

[Turn over

Marks | KU | PS

19. Roy wanted to show that chemicals can be used to produce an electric current.

ammeter   crocodile clips

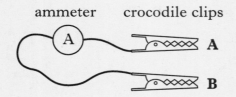

A

B

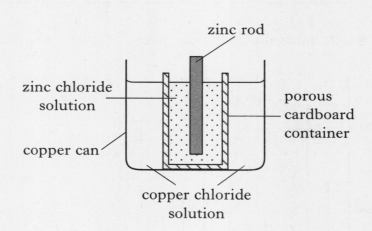

zinc rod

zinc chloride solution

copper can

copper chloride solution

porous cardboard container

When the crocodile clips (labelled **A** and **B**) were attached to certain parts of the apparatus, the ammeter gave a reading.

(a) (i) Show clearly **on the diagram**, using labels **A** and **B**, where the crocodile clips could have been attached.

1

(ii) Why was no current produced when the porous cardboard container was replaced by a glass beaker?

_____

_____

_____

1

(iii) What would happen to the reading on the ammeter if the zinc rod was replaced with a tin rod in a tin chloride solution?

_____

1

DO NOT WRITE IN THIS MARGIN

Marks | KU | PS

19. (continued)

(*b*) Roy was instructed to make 50 cm^3 of a 1 mol/litre solution of copper chloride, $CuCl_2$.

Calculate the mass, in grams, of copper chloride needed.

Show your working clearly

Answer: ______________ g **2**

(5)

[*END OF QUESTION PAPER*]

ADDITIONAL SPACE FOR ANSWERS

ADDITIONAL GRAPH PAPER FOR QUESTION 12(*a*)

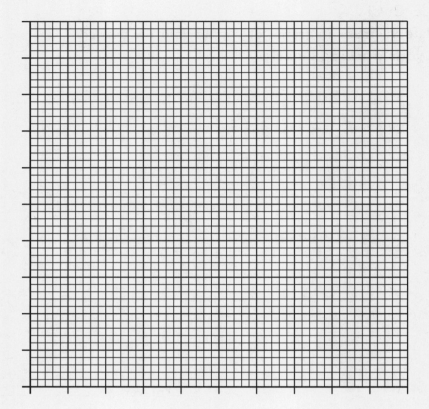

[BLANK PAGE]

C

FOR OFFICIAL USE

KU   PS

Total
Marks

0500/402

NATIONAL
QUALIFICATIONS
2004

MONDAY, 10 MAY
10.50 AM – 12.20 PM

CHEMISTRY
STANDARD GRADE
Credit Level

Fill in these boxes and read what is printed below.

Full name of centre

Town

Forename(s)

Surname

Date of birth

Day Month Year   Scottish candidate number   Number of seat

1 All questions should be attempted.

2 Necessary data will be found in the Data Booklet provided for Chemistry at Standard Grade and Intermediate 2.

3 The questions may be answered in any order but all answers are to be written in this answer book, and must be written clearly and legibly in ink.

4 Rough work, if any should be necessary, as well as the fair copy, is to be written in this book.

 Rough work should be scored through when the fair copy has been written.

5 Additional space for answers and rough work will be found at the end of the book.

6 The size of the space provided for an answer should not be taken as an indication of how much to write. It is not necessary to use all the space.

7 Before leaving the examination room you must give this book to the invigilator. If you do not, you may lose all the marks for this paper.

SCOTTISH
QUALIFICATIONS
AUTHORITY

©

PART 1

In **Questions 1 to 9** of this part of the paper, an answer is given by circling the appropriate letter (or letters) in the answer grid provided.

In some questions, two letters are required for full marks.

If more than the correct number of answers is given, marks will be deducted.

A total of 20 marks is available in this part of the paper.

SAMPLE QUESTION

A CH_4	B H_2	C CO_2
D CO	E C_2H_5OH	F C

(a) Identify the hydrocarbon.

Ⓐ	B	C
D	E	F

The correct answer to part (a) is A. This should be circled.

(b) Identify the **two** elements.

A	Ⓑ	C
D	E	Ⓕ

As indicated in this question, there are **two** correct answers to part (b). These are B and F. Both answers are circled.

If, after you have recorded your answer, you decide that you have made an error and wish to make a change, you should cancel the original answer and circle the answer you now consider to be correct. Thus, in part (a), if you want to change an answer A to an answer D, your answer sheet would look like this:

(A̶)	B	C
Ⓓ	E	F

If you want to change back to an answer which has already been scored out, you should enter a tick (✓) in the box of the answer of your choice, thus:

✓(A̶)	B	C
(D̶)	E	F

DO NOT WRITE IN THIS MARGIN

Marks | KU | PS

1. The grid shows the formulae of six oxides.

A H_2O	B NO_2	C K_2O
D CaO	E CO	F SO_2

(*a*) Identify the oxide produced by the sparking of air in car engines.

A	B	C
D	E	F

1

(*b*) Identify the **two** oxides produced by burning hydrocarbons.

A	B	C
D	E	F

1

(2)

[Turn over

DO NOT WRITE IN THIS MARGIN

Marks | KU | PS

2. The names of some hydrocarbons are shown in the grid.

A cyclobutane	B cyclopentane	C butane
D propane	E ethane	F butene

(*a*) Identify the hydrocarbon which is a liquid at 25 °C.

You may wish to use the data booklet to help you.

A	B	C
D	E	F

1

(*b*) Identify the **two** isomers.

A	B	C
D	E	F

1

(*c*) Identify the hydrocarbon that reacts quickly with bromine solution.

A	B	C
D	E	F

1

(3)

Marks | KU | PS

3. The grid shows the names of some soluble compounds.

A	B	C
magnesium bromide	sodium bromide	lithium hydroxide
D	**E**	**F**
sodium iodide	potassium sulphate	lithium chloride

(a) Identify the base.

A	B	C
D	E	F

1

(b) Identify the **two** compounds whose solutions would form a precipitate when mixed.

You may wish to use the data booklet to help you.

A	B	C
D	E	F

1

(c) Identify the compound with a formula of the type XY_2, where **X** is a metal.

A	B	C
D	E	F

1

(3)

[Turn over

4. A pupil carried out the following experiments.

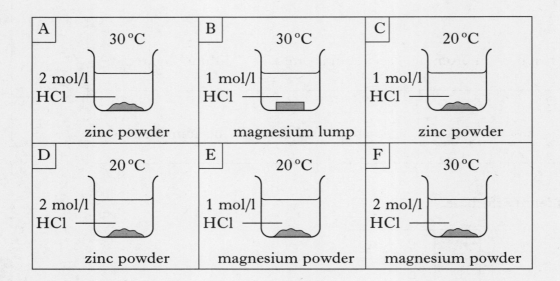

(a) Identify the **two** experiments which can be used to investigate the effect of concentration on the rate of the reaction.

A	B	C
D	E	F

(b) Identify the experiment with the fastest reaction rate.

A	B	C
D	E	F

1

1
(2)

Marks | KU | PS

5. The table contains information about some substances.

Substance	Melting point/°C	Boiling point/°C	Conducts as	
			a solid	a liquid
A	455	1567	no	yes
B	80	218	no	no
C	1492	2897	yes	yes
D	1407	2357	no	no
E	645	1287	no	yes
F	98	890	yes	yes

(a) Identify the **two** ionic compounds.

A
B
C
D
E
F

1

(b) Identify the substance which exists as a covalent network.

A
B
C
D
E
F

1

(2)

[Turn over

6.   The grid shows information about some particles.

A		B		C	
	$^{34}_{16}S^{2-}$		$^{24}_{12}Mg^{2+}$		$^{39}_{19}K$
D		E		F	
	$^{40}_{19}K$		$^{40}_{20}Ca$		$^{35}_{17}Cl^{-}$

(a) Identify the **two** particles which are isotopes.

A	B	C
D	E	F

1

(b) Identify the **two** particles with the same electron arrangement as argon.

A	B	C
D	E	F

1

(2)

DO NOT WRITE IN THIS MARGIN

Marks KU PS

7. The grid contains information about the particles found in atoms.

A charge = zero	B relative mass almost zero	C charge = 1–
D found inside the nucleus	E charge = 1+	F relative mass = 1

Identify the **two** terms which can be applied to electrons.

A	B	C
D	E	F

(2)

[Turn over

DO NOT WRITE IN THIS MARGIN

Marks | KU | PS

8. Glucose, sucrose and starch are carbohydrates.

Identify the **two** correct statements.

A	Glucose molecules join together with the loss of water.
B	Starch is a polymer made from sucrose molecules.
C	Sucrose turns warm Benedict's solution orange.
D	Glucose is an isomer of sucrose.
E	Starch dissolves easily in water.
F	Sucrose can be hydrolysed.

A
B
C
D
E
F

(2)

Marks | KU | PS

9. The diagram shows how an object can be coated in nickel.

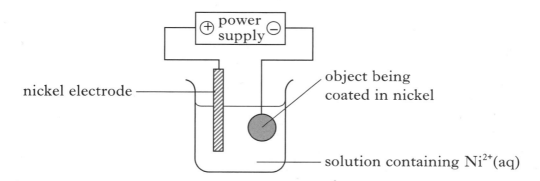

The following reactions take place at the electrodes.

Negative electrode: $Ni^{2+}(aq) + 2e^- \rightarrow Ni(s)$

Positive electrode: $Ni(s) \rightarrow Ni^{2+}(aq) + 2e^-$

Identify the **two** correct statements.

A	Nickel ions move towards the nickel electrode.
B	The mass of the nickel electrode decreases.
C	The process is an example of galvanising.
D	Oxidation occurs at the nickel electrode.
E	Electrons flow through the solution.

A
B
C
D
E

(2)

[Turn over

Marks | KU | PS

PART 2

A total of 40 marks is available in this part of the paper.

10. The structure of part of a polyacrylonitrile molecule is shown below.

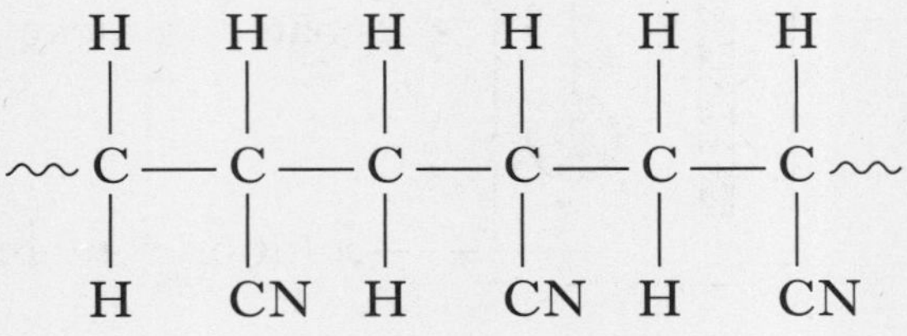

$$\sim C - C - C - C - C - C \sim$$

(with H, H, H, H, H, H on top of each carbon and H, CN, H, CN, H, CN on the bottom)

(a) Draw the structural formula for the monomer used to make polyacrylonitrile.

1

(b) Name a toxic gas produced when polyacrylonitrile burns.

_____

1

(2)

11. Some Euro coins are made from a hard-wearing alloy called Nordic Gold.

(a) What is an alloy?

_____

_____   1

(b) The composition of Nordic Gold is shown in the table.

Metal	copper	aluminium	zinc	tin
% by mass	89	5	5	1

One of the coins has a mass of 5·74 g.

(i) Calculate the mass, in grams, of aluminium in the coin.
Show your working clearly.

_____ g   1

(ii) Calculate the number of moles of aluminium in the coin.
Show your working clearly.

_____ mol   1

(3)

[Turn over

Marks | KU | PS

12. A group of pupils investigated the speed of reaction between marble chips (calcium carbonate) and hydrochloric acid, concentration 1 mol/l.

They used excess hydrochloric acid to make sure all the calcium carbonate had been used up.

The pupils used a balance to measure the mass lost during the reaction.

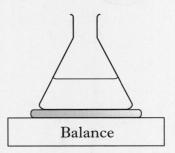

Balance

The results are shown in the table.

Time/minutes	0	0·5	1·0	2·0	3·0	4·0	5·0
Mass lost/g	0	0·30	0·50	0·70	0·76	0·79	0·80

(*a*) Why is mass lost during the reaction?

_____

_____   1

12. (continued)

(b) Draw a line graph of the results.

Use appropriate scales to fill most of the graph paper.

(Additional graph paper, if required, will be found on page 26.)

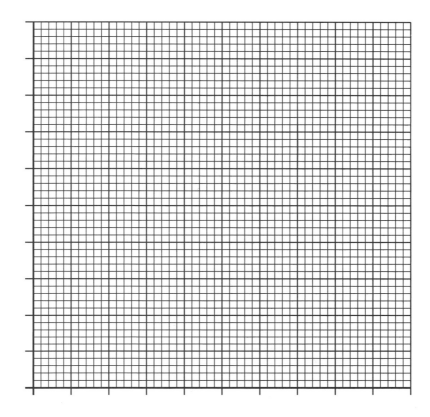

2

(c) The experiment was repeated using the same volume of hydrochloric acid, but with a concentration of 2 mol/l.

What mass loss would have been recorded?

_____ g

1

(d) Name the salt produced when marble chips react with hydrochloric acid.

_____

1

(5)

[Turn over

Marks | KU | PS

13. In a hydrogen molecule the atoms share two electrons in a covalent bond.

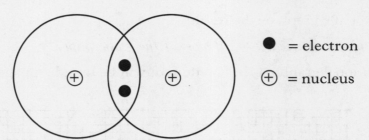

● = electron

⊕ = nucleus

(a) Explain how the covalent bond holds the two hydrogen atoms together.

_____

_____

_____

1

(b) The hydrogen molecule can be represented more simply as

H ⦂ H

(i) Showing **all** outer electrons, draw a similar diagram to represent a molecule of ammonia, NH_3.

1

(ii) Draw another diagram to show the **shape** of an ammonia molecule.

1

(3)

14. Methanol can take part in many chemical reactions.

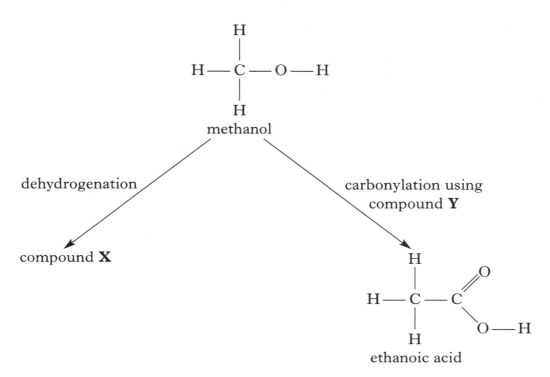

methanol

dehydrogenation

carbonylation using
compound **Y**

compound **X**

ethanoic acid

(a) (i) Compound **X** has the molecular formula CH_2O.
Draw the full structural formula for compound **X**.

1

(ii) Methanol is changed to compound **X** by dehydrogenation.
Suggest what is meant by **dehydrogenation**.

_____

_____

_____

1

(b) In carbonylation, methanol reacts with compound **Y** forming only
ethanoic acid.

Suggest a name for compound **Y**.

_____

1

(3)

Marks | KU | PS

15. Gemma and Laura set up the simple cell shown below.

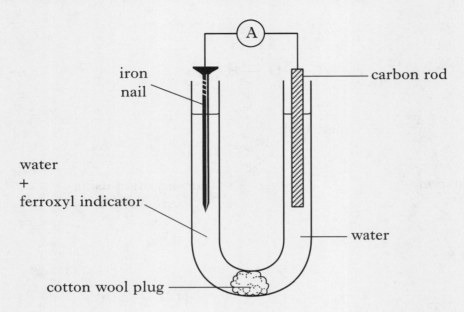

After two days the iron nail had rusted and the ferroxyl indicator had turned blue.

(*a*) **On the diagram**, clearly mark the path and direction of electron flow.

1

(*b*) The reaction taking place at the carbon rod produces hydroxide ions. How could Gemma and Laura have shown that hydroxide ions were present?

_____

_____

1

(2)

Marks | KU | PS

16. Electrolysis is a common industrial process. Some uses of electrolysis are shown in the diagram.

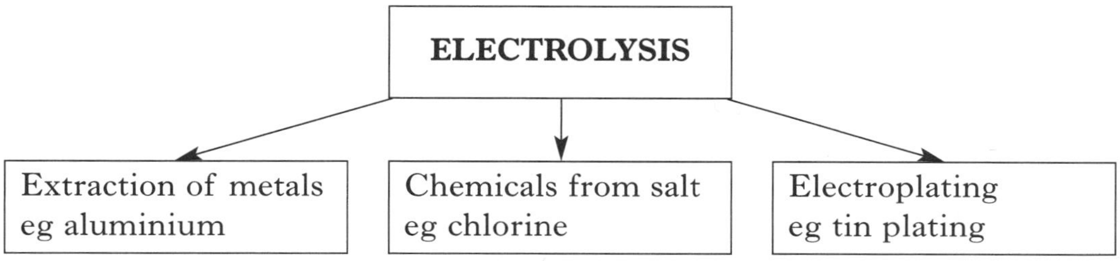

(a) State what is meant by electrolysis.

_____

_____

_____   1

(b) Aluminium is extracted by electrolysis of its molten oxide since aluminium oxide does not react when heated with carbon.

Why does aluminium oxide **not** react with hot carbon?

_____

_____

_____   1

(c) Chlorine is produced by the electrolysis of sodium chloride solution.
Write the ion-electron equation for the formation of chlorine.
You may wish to use the data booklet to help you.

_____   1

(d) Tin plated iron rusts very rapidly if the plating is scratched.
Explain why the iron rusts so rapidly.

_____

_____

_____   2

                                                          (5)

[Turn over

Marks

KU | PS

17. Silver jewellery slowly tarnishes in air. This is due to the formation of silver(I) sulphide, Ag_2S.

The silver(I) sulphide can be converted back to silver as follows.

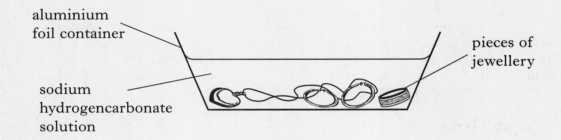

aluminium
foil container

pieces of
jewellery

sodium
hydrogencarbonate
solution

(a) Write the ionic formula for sodium hydrogencarbonate.
You may wish to use the data booklet to help you.

_____

1

(b) The equation for the reaction which takes place in the aluminium container is:

$$Ag_2S \quad + \quad Al \quad \rightarrow \quad Ag \quad + \quad Al_2S_3$$

(i) Balance this equation.

1

(ii) Name the type of chemical reaction which takes place.

_____

1

(c) Calculate the percentage by mass of aluminium in Al_2S_3.
Show your working clearly.

2

(5)

Marks | KU | PS

18. Fritz was investigating the properties of ammonia.

Before   **After**

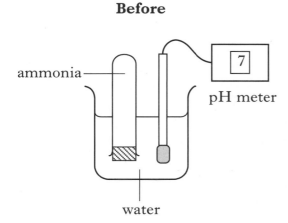 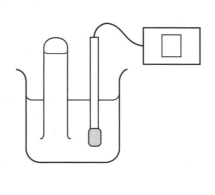

(a) Why did the water rise up the test tube when the stopper was removed?

_____

_____   1

(b) When the stopper was removed the reading on the pH meter changed. Suggest what the new reading would have been.

_____   1

                                         (2)

[Turn over

Official SQA Past Papers: Credit Chemistry 2004

DO NOT
WRITE IN
THIS
MARGIN

Marks | KU | PS

19. Water can exist in three different states: solid, liquid and gas.

The state depends on the temperature and pressure.

The diagram below shows these relationships.

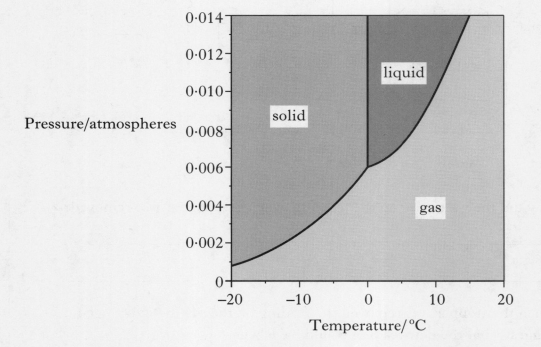

(a) In which state would water exist at 15 °C and 0·007 atmospheres?

_____

1

(b) Solid water at 0·004 atmospheres is allowed to warm up. The pressure is kept constant.

At what temperature would the solid water change into a gas?

_____ °C

1

(2)

Marks | KU | PS

20. Dienes are a homologous series of hydrocarbons which contain two double bonds per molecule.

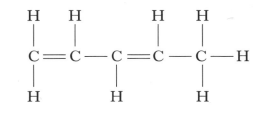

buta-1,3-diene   penta-1,3-diene

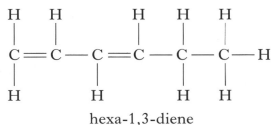

hexa-1,3-diene

(a) What is meant by the term "homologous series"?

_____

_____

_____   1

(b) Suggest a general formula for the dienes.

_____   1

(c) Write the **molecular formula** for the product of the complete reaction of penta-1,3-diene with bromine.

_____   1

(d) Draw a full structural formula for an isomer of buta-1,3-diene which contains only **one** double bond per molecule.

1

(4)

21. A pupil carried out a titration using the chemicals and apparatus shown below.

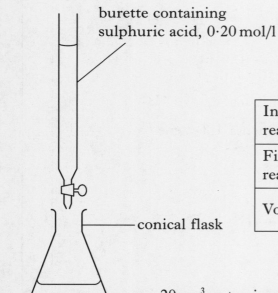

burette containing
sulphuric acid, 0·20 mol/l

conical flask

— 20 cm³ potassium hydroxide solution + indicator

	Rough titre	1st titre	2nd titre
Initial burette reading/cm³	0·5	21·7	0·3
Final burette reading/cm³	21·7	42·4	20·8
Volume used/cm³	21·2	20·7	20·5

(a) How would the pupil know when to stop adding acid from the burette?

_____

_____   **1**

(b) (i) What average volume should be used to calculate the number of moles of sulphuric acid needed to neutralise the potassium hydroxide solution?

_____ cm³   **1**

DO NOT WRITE IN THIS MARGIN

Marks | KU | PS

21. (b) (continued)

(ii) Calculate the number of moles of sulphuric acid in this average volume.

Show your working clearly.

__________ mol **1**

(iii) The equation for the titration reaction is

$$H_2SO_4 \quad + \quad 2KOH \quad \rightarrow \quad K_2SO_4 \quad + \quad 2H_2O$$

Calculate the number of moles of potassium hydroxide in 20 cm^3 of the potassium hydroxide solution.

Show your working clearly.

__________ mol **1**

(4)

[*END OF QUESTION PAPER*]

ADDITIONAL SPACE FOR ANSWERS

ADDITIONAL GRAPH PAPER FOR QUESTION 12(*b*)

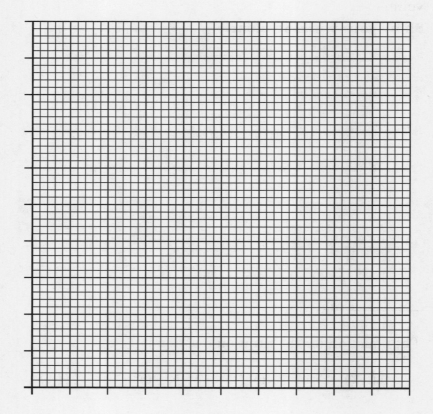

BLANK PAGE

FOR OFFICIAL USE

C

	KU	PS
Total Marks		

0500/402

NATIONAL
QUALIFICATIONS
2005

MONDAY, 9 MAY
10.50 AM – 12.20 PM

CHEMISTRY
STANDARD GRADE
Credit Level

Fill in these boxes and read what is printed below.

Full name of centre

Town

Forename(s)

Surname

Date of birth
Day Month Year   Scottish candidate number   Number of seat

1   All questions should be attempted.

2   Necessary data will be found in the Data Booklet provided for Chemistry at Standard Grade and Intermediate 2.

3   The questions may be answered in any order but all answers are to be written in this answer book, and must be written clearly and legibly in ink.

4   Rough work, if any should be necessary, as well as the fair copy, is to be written in this book.

Rough work should be scored through when the fair copy has been written.

5   Additional space for answers and rough work will be found at the end of the book.

6   The size of the space provided for an answer should not be taken as an indication of how much to write.   It is not necessary to use all the space.

7   Before leaving the examination room you must give this book to the invigilator.   If you do not, you may lose all the marks for this paper.

SCOTTISH
QUALIFICATIONS
AUTHORITY

©

PART 1

In Questions 1 to 9 of this part of the paper, an answer is given by circling the appropriate letter (or letters) in the answer grid provided.

In some questions, two letters are required for full marks.

If more than the correct number of answers is given, marks will be deducted.

A total of 20 marks is available in this part of the paper.

SAMPLE QUESTION

A CH_4	B H_2	C CO_2
D CO	E C_2H_5OH	F C

(a) Identify the hydrocarbon.

Ⓐ	B	C
D	E	F

The one correct answer to part (a) is A. This should be circled.

(b) Identify the **two** elements.

A	Ⓑ	C
D	E	Ⓕ

As indicated in this question, there are **two** correct answers to part (b). These are B and F. Both answers are circled.

If, after you have recorded your answer, you decide that you have made an error and wish to make a change, you should cancel the original answer and circle the answer you now consider to be correct. Thus, in part (a), if you want to change an answer A to an answer D, your answer sheet would look like this:

A̷	B	C
Ⓓ	E	F

If you want to change back to an answer which has already been scored out, you should enter a tick (✓) in the box of the answer of your choice, thus:

✓A̷	B	C
D̷	E	F

Marks | KU | PS

1. The grid shows the names of some elements.

A	B	C
argon	potassium	magnesium
D	E	F
chlorine	phosphorus	sulphur

(*a*) Identify the element which produces a lilac flame colour.

You may wish to use the data booklet to help you.

A	B	C
D	E	F

1

(*b*) Identify the element with atoms which have the same electron arrangement as a Ca^{2+} ion.

A	B	C
D	E	F

1

(*c*) Identify the **two** elements which would form a covalent compound with a formula of the type X_3Y_2.

A	B	C
D	E	F

1

(3)

[Turn over

2. Distillation of crude oil produces several fractions.

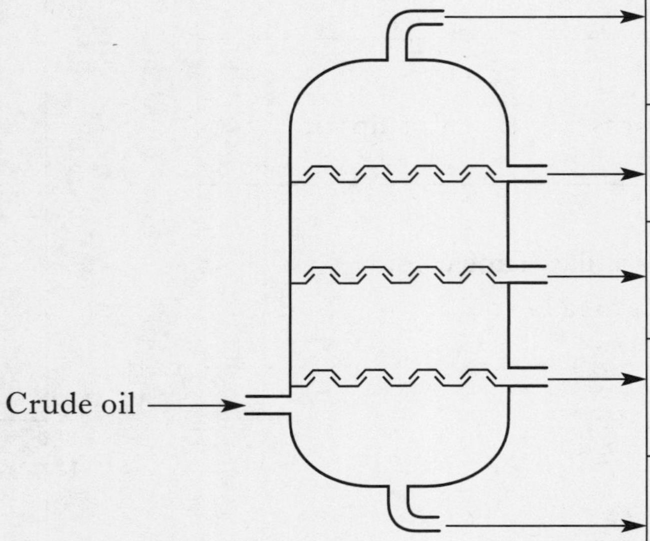

Fraction	Number of carbon atoms per molecule
A	1–4
B	4–10
C	10–16
D	16–20
E	20+

Crude oil ⟶

(a) Identify the fraction which is used as a fuel in camping gas stoves.

A
B
C
D
E

1

(b) Identify the fraction with the highest viscosity.

A
B
C
D
E

1

(2)

Marks KU PS

3. The names of some carbohydrates are shown.

A	glucose
B	fructose
C	maltose
D	sucrose
E	starch

(a) Identify the carbohydrate which does not dissolve well in water.

| A |
| B |
| C |
| D |
| E |

1

(b) Identify the **two** carbohydrates which are disaccharides.

| A |
| B |
| C |
| D |
| E |

1

(2)

[Turn over

Page five

4. The table contains information about some substances.

Substance	Melting point/°C	Boiling point/°C	Conducts as a solid	Conducts as a liquid
A	−7	59	no	no
B	98	883	yes	yes
C	−39	357	yes	yes
D	547	1265	no	yes
E	−78	−33	no	no
F	1700	2230	no	no

(a) Identify the substance which is a gas at 0 °C.

A
B
C
D
E
F

1

(b) Identify the **two** substances which exist as molecules.

A
B
C
D
E
F

1

(2)

Marks | KU | PS

5. The grid shows the formulae of some hydrocarbons.

A	B	C
$CH_3 - CH - CH_3$ 	 CH_3	CH_2 ╱ ╲ $CH_2 - CH_2$	$CH_3 - CH_2 - CH_2 - CH_3$
D	E	F

CH_3
     ╲
      $C = CH_2$
     ╱
 CH_3
	$CH_2 - CH_2$
	
			 $CH_2 - CH_2$
				CH_3 H
				   ╲ ╱
				    $C = C$
				   ╱ ╲
				  H H

(a) Identify the hydrocarbon which can be used to make poly(propene).

A	B	C
D	E	F

1

(b) Identify the **two** hydrocarbons with the general formula C_nH_{2n} which do **not** react quickly with bromine solution.

A	B	C
D	E	F

1

(c) Identify the **two** isomers of

$$CH_2 = CH - CH_2 - CH_3$$

A	B	C
D	E	F

2

(4)

[Turn over

6. There are many different types of chemical reaction.

A	B	C
reduction	precipitation	displacement
D	E	F
hydrolysis	neutralisation	oxidation

Identify the following type of reaction.

$$SO_3^{2-}(aq) \ + \ H_2O(\ell) \ \longrightarrow \ SO_4^{2-}(aq) \ + \ 2H^+(aq) \ + \ 2e^-$$

A	B	C
D	E	F

(1)

DO NOT WRITE IN THIS MARGIN

Marks KU PS

7. The grid shows some statements which could be applied to a solution.

A	It does not react with magnesium.
B	It has a pH less than 7.
C	It does not conduct electricity.
D	It produces chlorine gas when electrolysed.
E	It contains more H^+ ions than pure water.

Identify the **two** statements which are true for **both** dilute hydrochloric acid and dilute sulphuric acid.

A
B
C
D
E

(2)

[Turn over

Marks | KU | PS

8. David was studying the reactions of some metals and their compounds.

He carried out experiments involving magnesium, copper, zinc, nickel, silver and an unknown metal **X**.

Listed below are some of the observations he recorded.

A	**X** was more readily oxidised than copper.
B	**X** oxide was more stable to heat than silver oxide.
C	Magnesium displaced **X** from a solution of **X** nitrate.
D	**X** reacted more vigorously than nickel with dilute acid.
E	Compounds of **X** were more readily reduced than compounds of zinc.

From his observations, David produced the following order of reactivity.

magnesium, zinc, nickel, copper, **X**, silver

———————————————————→

decreasing activity

Identify the **two** observations which can be used to show that **X** has been wrongly placed.

A
B
C
D
E

(2)

9. Equations are used to represent chemical reactions.

A	$H^+(aq) + OH^-(aq) \rightarrow H_2O(\ell)$
B	$Fe^{3+}(aq) + e^- \rightarrow Fe^{2+}(aq)$
C	$Fe(s) \rightarrow Fe^{2+}(aq) + 2e^-$
D	$Fe^{2+}(aq) + 2e^- \rightarrow Fe(s)$
E	$H_2(g) \rightarrow 2H^+(aq) + 2e^-$
F	$2H_2O(\ell) + O_2(g) + 4e^- \rightarrow 4OH^-(aq)$

Identify the **two** equations which are involved in the corrosion of iron.

| A |
| B |
| C |
| D |
| E |
| F |

(2)

[Turn over

[Turn over for Question 10 on *Page thirteen*

Page twelve

Marks | KU | PS

PART 2

A total of 40 marks is available in this part of the paper.

10. On some boats the steel propellers have zinc blocks attached to help prevent rusting. The zinc is oxidised, protecting the steel.

zinc block   steel propeller

  (a) (i) Write the ion-electron equation for the oxidation of zinc.
You may wish to use the data booklet to help you.

_____ **1**

       (ii) What name is given to the **type** of protection provided by the zinc?

_____ **1**

  (b) If cobalt is used instead of zinc the steel propeller rusts quickly.
What does this suggest about the reactivity of cobalt compared to iron?

_____

_____ **1**

                                       (3)

[Turn over

11. Polystyrene is an addition polymer. It is made from the monomer styrene.

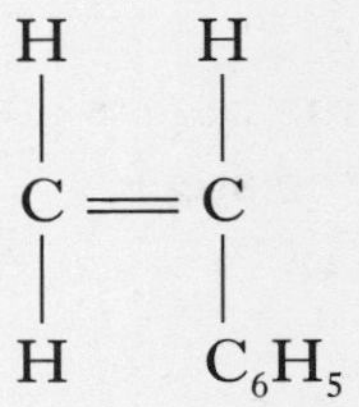

$$\begin{array}{ccc} H & & H \\ | & & | \\ C & = & C \\ | & & | \\ H & & C_6H_5 \end{array}$$

styrene

(a) Draw a section of the polystyrene structure, showing three monomer units joined together.

1

(b) Calculate the percentage by mass of carbon in a molecule of styrene.

Answer _____ %   2

(3)

Marks | KU | PS

12. Methane (CH_4), ethane (C_2H_6) and propane (C_3H_8) are the first three members of the alkanes.

(a) State the general formula for the alkanes.

_____

1

(b) The ninth member of the alkanes is nonane (C_9H_{20}).

   (i) Predict the boiling point of nonane.
      You may wish to use page 6 of the data booklet to help you.

      _____ °C

1

   (ii) Nonane can be cracked to produce smaller, more useful hydrocarbons. A catalyst is used to speed up this reaction.
      Suggest another reason for using a catalyst.

      _____

      _____

1

(c) Alkanes can be made by the reaction of sodium with iodoalkanes.
For example, butane can be made from iodoethane.

```
   H  H                    H  H                    H   H   H   H
   |  |                    |  |                    |   |   |   |
H−C−C−I  +  2Na  +  I−C−C−H   →   H−C−C−C−C−H  +  2NaI
   |  |                    |  |                    |   |   |   |
   H  H                    H  H                    H   H   H   H
  iodoethane             iodoethane                   butane
```

Butane can also be made using two **different** iodoalkanes.
Name the **two** iodoalkanes.

_____

_____

1

(4)

[Turn over

13. A mass spectrometer is an instrument that can be used to measure the percentage of isotopes in a sample of an element.

When a sample of chlorine is passed through a mass spectrometer the following graph is obtained.

Each spike on the graph shows the presence of an isotope.

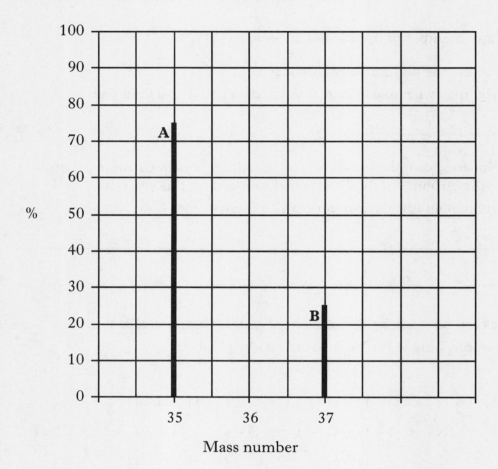

The **relative atomic mass** of an element can be calculated using the formula:

$$\frac{(\text{mass of isotope } \mathbf{A} \times \%) + (\text{mass of isotope } \mathbf{B} \times \%)}{100}$$

The **relative atomic mass** of chlorine $= \dfrac{(35 \times 75) + (37 \times 25)}{100}$

$= 35 \cdot 5$

Official SQA Past Papers: Credit Chemistry 2005

Marks

DO NOT
WRITE IN
THIS
MARGIN

KU   PS

13. (continued)

(a) The following graph was obtained for a sample of lithium.

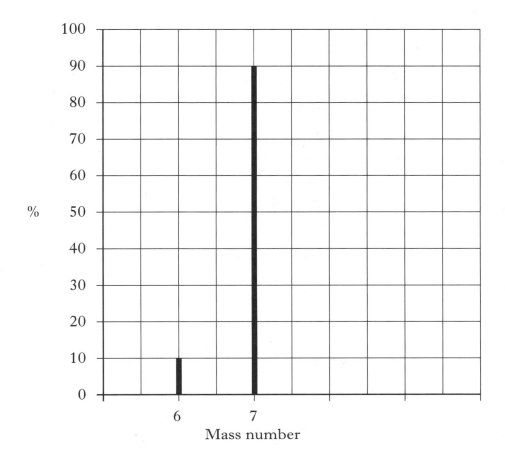

Mass number

(i) How many isotopes are present in the sample of lithium?

_____

1

(ii) Using the information in the graph calculate the relative atomic mass of lithium.

Show your working clearly.

1

(b) Complete the table to show the number of each type of particle in the ion, $_3^7Li^+$.

Particle	Number
proton	
neutron	
electron	

2

(4)

Marks

14. Clare carried out an experiment to make copper chloride crystals.

Instructions for preparation of copper chloride crystals

Step 1 Add $25\,cm^3$ of dilute hydrochloric acid to a beaker.

Step 2 Add a spatulaful of copper carbonate powder to the acid and stir.

Step 3 Continue adding copper carbonate until some of the solid remains.

Step 4

Step 5

(*a*) Why did Clare continue to add copper carbonate until some solid remained?

_____

_____   1

(*b*) Name the **two** techniques which Clare would have carried out in steps **4** and **5** to prepare a sample of copper chloride crystals.

Step 4 _____

Step 5 _____   2

(3)

Marks | KU | PS

15. Two atoms of nitrogen share electrons to form a nitrogen molecule.

(a) Draw a diagram to show how the outer electrons are arranged in a molecule of nitrogen, N_2.

1

(b) Oxides of nitrogen dissolve in water to produce nitric acid.

(i) Name the industrial process used to manufacture nitric acid.

_____

1

(ii) A platinum catalyst is used in the industrial manufacture of nitric acid.

Why is it **not** necessary to continue heating the platinum once the reaction has started?

_____

_____

1
(3)

[Turn over

Marks KU PS

16. Dilute hydrochloric acid reacts with sodium thiosulphate solution ($Na_2S_2O_3$) to produce a precipitate of sulphur.

$$HCl + Na_2S_2O_3 \rightarrow NaCl + S + SO_2 + H_2O$$

(a) Balance this equation.

1

(b) A pupil investigated the effect of temperature on the speed of the reaction. She measured the time taken for enough sulphur to form to make the cross disappear.

cross visible   cross not visible

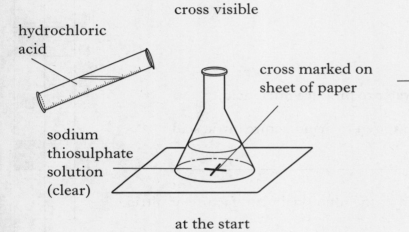
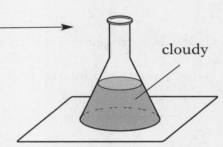

hydrochloric acid

cross marked on sheet of paper

cloudy

sodium thiosulphate solution (clear)

at the start   later

Her results are shown in the table.

Temperature/°C	Time/s
25	89
30	64
35	44
40	33
45	27
50	21

16.   (b) (continued)

(i)   Draw a line graph of the results.

Use appropriate scales to fill most of the graph paper.

(Additional graph paper, if required, will be found on page 27.)

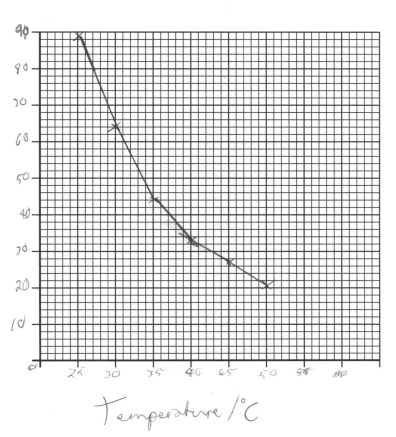

2

(ii)   Use your graph to estimate the time taken, in seconds, for the cross to disappear at 60 °C.

_____

1

(iii)   Describe the relationship between the temperature and the **speed** of the reaction.

_____

_____

1

(c)   State **one** factor that must be kept constant throughout this investigation.

_____

1

(6)

[Turn over

Marks   KU   PS

17. Glucose and starch are both carbohydrates.

(*a*) Write the molecular formula for glucose.

_____ 1

(*b*) A pupil set up the following experiment to turn starch into glucose.

Test tube **A**

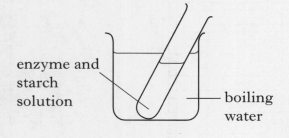

enzyme and
starch
solution
— boiling
water

Test tube **B**

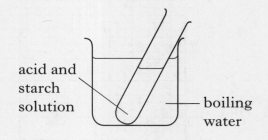

acid and
starch
solution
— boiling
water

(i) Name the type of chemical reaction which takes place when starch is broken down to glucose.

_____ 1

(ii) Suggest why glucose would **not** be formed in test tube **A**.

_____

_____

_____ 1

(3)

Marks | KU | PS

18. Copper can be extracted from its oxide by heating copper(II) oxide with hydrogen gas. Water is also formed during the reaction.

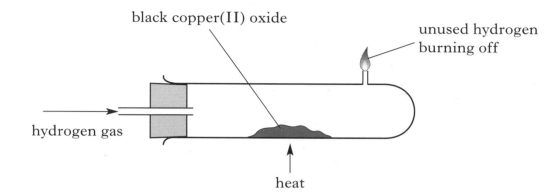

black copper(II) oxide

unused hydrogen burning off

hydrogen gas

heat

(a) Write an equation, using symbols and formulae, for the reaction between copper(II) oxide and hydrogen gas.
There is no need to balance the equation.

_____

1

(b) Suggest the colour change which would be seen in the copper(II) oxide during the reaction.

_____

1

(c) Suggest why calcium cannot be extracted from its oxide by heating with hydrogen gas.

_____

_____

_____

1

(3)

[Turn over

19. The cell below can be used in a carbon monoxide detector.

Carbon monoxide enters the cell along with oxygen from the air at electrode **A**.

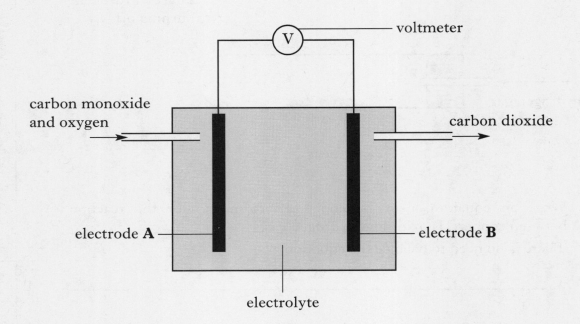

The reactions occurring at each electrode are:

Electrode A

$$CO(g) + H_2O(\ell) \longrightarrow CO_2(g) + 2H^+(aq) + 2e^-$$

Electrode B

$$2H^+(aq) + \tfrac{1}{2}O_2(g) + 2e^- \longrightarrow H_2O(\ell)$$

(a) **On the diagram**, clearly mark the path and direction of electron flow.

1

(b) What is the purpose of the electrolyte in the above cell?

_____

_____

1

Marks | KU | PS

19. **(continued)**

(c) Sugar solution cannot be used as an electrolyte.

What does this indicate about the bonding in sugar?

_____

_____   1

(d) Platinum is used for the electrodes in this cell.

(i) To which family of metals does platinum belong?

_____   1

(ii) Platinum is also used as a catalyst in a catalytic converter in car exhausts.

What does a catalytic converter do?

_____

_____

_____   1

                                                  (5)

[Turn over

Marks | KU | PS

20. Aluminium powder reacts with dilute sulphuric acid.

$$2Al(s) \quad + \quad 3H_2SO_4(aq) \quad \longrightarrow \quad Al_2(SO_4)_3(aq) \quad + \quad 3H_2(g)$$

(a) Circle the formula for the salt in the above equation.

1

(b) Calculate the mass of hydrogen produced when 1·35 g of aluminium reacts with dilute sulphuric acid.

Answer _____ g   **2**

(3)

[END OF QUESTION PAPER]

ADDITIONAL SPACE FOR ANSWERS

ADDITIONAL GRAPH PAPER FOR QUESTION 16(*b*)(i)

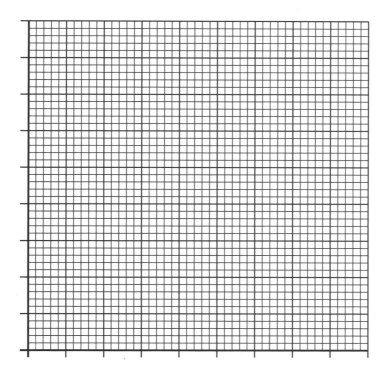

DO NOT
WRITE IN
THIS
MARGIN

KU	PS

ADDITIONAL SPACE FOR ANSWERS

[BLANK PAGE]

[BLANK PAGE]

[BLANK PAGE]

[BLANK PAGE]